AVOID SILLY MISTAKES IN MATHEMATICS

Rajesh Sarswat

First Impression: 2017

Table of Contents

AMAZON BESTSELLER
(www.amazon.com)
(Kindle e-book Category)

ellers
n sales. Updated hourly.

Best Sellers in **Popular & Elementary Arithmetic**

Top 100 Paid   **Top 100 Free**

1. kindleunlimited

Avoid Silly Mistakes in...
› Rajesh Sarswat
3
Kindle Edition
$0.76

2.

Secrets of Mental Math:
› Arthur Benjamin
423
Kindle Edition

3. kindleunlimited

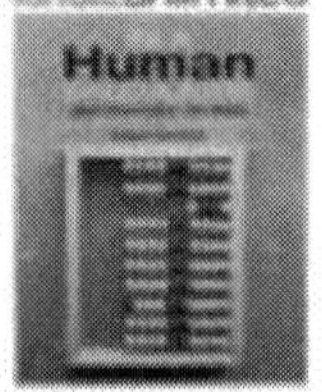

Be a Human Calculator:...
› Rajesh Sarswat

4. kindleunlimited

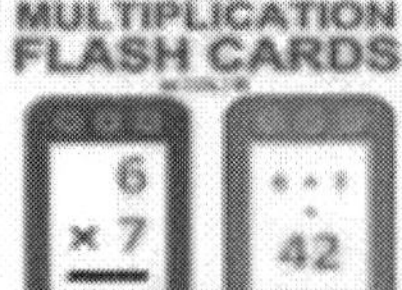

Digital Multiplication...
› Chris McMullen

Acknowledgements

I dedicate my fourth book to all of my students for teaching me the possible areas where students are prone to making silly mistakes.

Rajesh Sarswat

About the book

As a mathematics teacher, each year I am reminded just how crucial it is for all of my students to be familiar with the core elements of the topic.

To bring my students to the same level of understanding, I start the first three to four classes of every batch of Grade XI and Grade XII by teaching the primary topics of mathematics. These issues pertain to the early years of schooling and, focus on where the students may make mistakes.

In these classes, it is common for even the most gifted students to be committing silly mistakes. It is the case that, year after year, students are confused due to their weak fundamental understanding of the core topics.

What is interesting is that students are acutely aware of the numerous silly mistakes they make, but have no strategy to fix these problems.

This book aims to provide students with the action plan they require to identify and fix the problems they have with core mathematical concepts.

This book is not mathematics textbook and will not teach students more advanced topics such as calculus, vector, and matrices. Instead, the book will assume that students already have a basic knowledge of these issues and are struggling due to silly mistakes made in more fundamental concepts. The students can use this book alongside other more advanced textbooks as a way to re-enforce a student's knowledge.

My experience has shown me that while students may have a good grasp of more advanced topics; it is not unusual for them to have misconceptions of essential principles. In these cases, I have taken steps to provide a detailed and easy to understand-explanations that will help remove these mistakes.

This book will provide students, teachers, and parents with easy to understand explanations that help them to avoid silly mistakes in mathematics.

For any suggestions or any doubt about the topics covered in this book, readers may contact me at the following address. I will be glad to aid the readers to the best of my capacity.

Facebook Page:

https://www.facebook.com/avoidsillymistakes/

Email: rsarswat.rs@gmail.com

BY THE SAME AUTHOR

HOW TO MEMORIZE
FORMULAS IN
MATHEMATICS

BOOK – 1

CALCULUS

RAJESH SARSWAT

1
Types of Silly Mistakes

Learning math is not only about strong fundamentals but also about a lot of practice, and making mistakes is part of that process. In my opinion, making errors in math is a universal phenomenon, and can help the students to learn and explore math in a better way. However, repeating the same mistakes over an extended period will not benefit the students and will be harmful to their confidence. Hence, the primary goal of the parents and teachers is to help the students to learn to manage and overcome their habit of making silly mistakes.

There are different types of silly mistakes that students make.

It is essential that students are aware of the types of mistakes they are making. Once they understand their mistakes, they can try to learn from the techniques explained in this book to cut-down and rectify these mistakes.

General Techniques to Reduce Silly Mistakes

1. Slow Down a Bit

Students are often in a hurry to complete the questions they are attempting. A temptation to move to the next question, either due to anxiety or, even, overconfidence is the underlying reason to commit silly mistakes.

Students are advised to attempt the questions a bit slower and thus paying full attention to every step they're writing.

2. Analyzing the Errors/Mistakes

Students who want to reduce their stupid mistakes should keep a record of the types of errors they make during examinations or practice sessions. They should keep records maintaining the details of silly mistakes committed. The students may design the format of the records sheet as per their requirements. However, it should include the following information:

- Date of committing an error.
- Type of mistakes: careless, computational, conceptual, reading error, sloppy handwriting, copying a question wrong, missed the unit, etc.
- A summary of the question.
- Incorrect step(s) of the solution.
- Correct step(s) of the solution.
- The previous occasion of occurrence of such mistake (write the date).

At the outset, it looks awkward to keep up such a record, but only after practicing this method for a while; students will be able to understand the effectiveness of this technique.

Going through the records of the mistakes committed will leave a visual imprint of the error on the mind of the student. This process of reconciliation makes the student aware of potential mistakes each time they start a new question. Over time, the student will begin to pre-empt the errors and change the way they answer questions to avoid these silly mistakes.

3. Breath

How can breathing reduce stupid mistakes?

Shallow breathing leads to tension and stress in the body and thus reduces the clarity of thought. Whenever we are frightened or anxious, such as in examinations, our breathing slows down.

The solution is simple. Breathe deeply and slowly.

Before the start of every question take three slow and deep breaths to reduce the anxiety and stress levels.

4. Highlighting Important Information

When reading a problem, you should be highlighting important information (by circling or by underlining). This step will help you to understand problems better and reduce the number of silly mistakes.

5. Improve Handwriting

If an examiner is unable to understand the answer the student has written, then the answer will be marked as incorrect. Thus, due to poor handwriting, the students can be penalized, despite knowing the correct answer.

Students should, therefore, be encouraged to improve their handwriting, so that it is neat and legible. Thus the students have to work slowly and print their answers (rather than write in 'joined up' writing).

6. Stay Well Hydrated

It is a well-established scientific fact that the human brain consists mostly of water (around 80%). Students should drink adequate water before engaging in any mental activity. A glass of water before starting a homework session or test is helpful to keep the mind fit and thus reduces chances of silly mistakes in an indirect way.

7. Take Movement Breaks

As action helps the mental and creative processes, taking regular breaks is essential to keep the brain alert.

Possible break activities include:

1. Stretching the body even in sitting posture.
2. Walking around for some seconds.
3. Closing the eyes and breathing deeply for few seconds.

8. Have a Good Night Sleep

For well-prepared students, the night before the exam shouldn't be too stressful. In fact, such students should relax a bit so that they may have time to get themselves organized for the big day. Studies have found that if a person remained awake for 21 hours straight, he might have the mental state of a drunken person regarding his ability to concentrate, memorize and recall information, etc.

Students can't afford to stay awake all night studying for an exam because they just won't be active on the day of the exam. Students should make sure that they get on average 8 hours of sleep a night. By doing so, they will be mentally fit to take the examination and their fitness level will enable them to concentrate more and will thus cut-down chances of silly mistakes in an indirect way.

9. Eat Light

Even if students regularly skip breakfast or avoid eating when they are nervous on a test day, they should still make the time to eat something. A brain needs the energy from food to work efficiently. Students need to keep their mental focus on the exam and not on their hunger. If students really cannot stomach food, then try having a protein shake or something light.

Students should eat enough to feel satisfied but not so much as to sense full. If they eat a big breakfast or lunch before an exam, they will feel drowsy and dull, and their body's energy will be diverted to the digestive process whereas it should be provided to your brain to act efficiently.

10.Time Management

Students should try to complete the question paper at least 10 minutes before the scheduled end time. During this period, students may re-check all the tedious calculations and steps of those problems where they may commit silly mistakes as per analysis record maintained by them.

11. Motivation from Parents and Teachers

Every kid is unique in some way or the other. Even if some children are not getting good grades in mathematics due to various reasons, constant scolding and pressure techniques adopted by teachers and parents will not only de-motivate the kids but also block the chances of their further improvement.

If parents, teachers, and students try to work together on the general techniques specified in this book, there are high chances that the students will be able to reduce the quantum of silly mistakes committed by them at various levels to a minimal and it will be a win-win situation for all.

Types of Silly Mistakes

As per my interactions with students of different age groups during last 25 years of my teaching career, I've categorized the kinds of silly mistakes into three broad categories:

1. Casual or Careless Silly Mistakes

2. Calculation Errors

3. Conceptual Silly Mistakes

Let us now discuss each of these types in detail:

1. Casual or Careless Silly Mistakes

Casual or Careless Silly Mistakes occur due to lack of concentration, or hasty working on the part of students. Out of the three types of errors, this is the most common type and is carried out by most of the students often across the globe, at least at some point in time during their school days. However, these mistakes are most natural to correct, and students may rectify these errors with slight of focus and observation. Some examples of these silly errors may include:

(i) Copying the problem wrong

(ii) Copying the number wrong from the previous step

(iii) Reading/understanding the question wrong

(iv) Writing the answer without units

(v) Not marking and labeling the axes while doing questions based on graphs

(vi) Not writing the scales used for different axes while doing questions based on graphs

(vii) Dropping a negative sign

(viii) Sloppy handwriting

(ix) Not following the directions/instructions written before the question paper/question

(x) Incorrect rounding off

(xi) Leaving some questions unanswered

(xii) Using inappropriate brackets

(xiii) Multiplying in place of adding

Preventing Casual or Careless Silly Mistakes

Apart from following the general techniques to avoid silly mistakes, the following methods will also be beneficial for students for taking care of Careless Silly Mistakes committed by them:

(i)Leaving the Questions Unanswered

There is a tendency for students to leave questions unanswered where they have no clue. However, it is advisable that they should try to write whatever fact they can recollect the problem. It will always be better to write something and not leaving the answer sheet blank. There is a possibility that they may be awarded some marks in some of the questions answered by them reluctantly and that may result in improvement in their grades.

A similar method may be adopted for Multiple Choice question/ True-False Type question or Fill in the blank type questions. However, if there is negative marking for a wrong answer, it will be wiser to leave questions unanswered, where one had no clues.

(ii) Simplified Answer

Students are in such a hurry that they leave their answers without simplifying the answers.

Let us see some examples of an incomplete answer:

Example 1: Answer is 12/16

Mistake Committed: Here, the answer is a fraction and as per mathematical convention, students should reduce any fraction in its lowest form. So, the correct way to write the answer is 3/4.

Example 2: Answer is $x^2 + 5 + 7x + 2$

Mistake Committed: Here, the answer has two constants. A polynomial should have only one constant term. So, the correct way to write the answer is $x^2 + 7x + 7$.

Example 3: Answer is $x^2 + 5 + x^3 + 2x$

Mistake Committed: The student has not expressed the polynomial in the standard form. As per mathematical convention, the correct way to write the polynomials is in the decreasing order of exponents. So, the correct way of writing the answer is $x^3 + x^2 + 2x + 5$.

Example 4: Answer is 7/5

Mistake Committed: Here, the answer is an improper fraction and while writing solution such fractions should be expressed in mixed fraction form to give a clear picture. Thus, the correct way of writing the answer is $1\frac{2}{5}$.

(iii) Read the Instructions Carefully

This case is another example of a careless mistake by the students. Attempting a question without reading the instructions provided before the problem, may sometimes be so disastrous that it may wipe out all their marks.

See the following Example:

Example 1: Suppose in a problem, students have to mark the options which are not true.

Now if a student without going through the instructions carefully wrote all the answers by assuming that he has to answer what is right? Apparently, all his answers will be wrong even if he knew all the answers.

Example 2: If in a question, the directions were to round off the answer up to two places of decimals.

If the student, ignoring the instructions, wrote the answer as 2.3769, he will lose marks.

Example 3: If in a question of mensuration, all the units are in cm and the instructions are to find the volume in m^3.

As the units were in cm, the student, ignoring the instructions wrote the answer in cm^3 and so will lose marks.

(iv) Writing the Units Correctly

See the following answers given by a student:

Q.1: Area = 6

Q.2: Age = 7

Q 3: Volume = 3

In none of the question, the student has mentioned the unit of the answer. Students should write the answers along with the appropriate units as per the directions/data provided in the question.

(v) Show Full Steps

Students have a general tendency to skip few steps of the solution to arrive at the answer as early as possible. This tendency may be due to carelessness or for saving a little time during the examination.

Skipping steps is a common reason to commit silly mistakes by students as due to the omission of few steps, students are unable to see the error committed by them during the process.

Students are, therefore, suggested to write every step of the solution (along with the formulae used) neatly and cleanly and off-course, in legible handwriting. This simple advice will not only help them in reducing silly mistakes but also aid them to re-check the solution at the end of question paper in a proper fashion.

2. Calculation Errors

The second type of silly mistake is relating to calculations. It means that somewhere in the process of solving the problem, the student has done some wrong calculations.

Making a single computational mistake in a multi-step problem will make the rest of their work as, incorrect and that will make the last answer as wrong.

The main reason for committing such mistakes is poor calculation skills and casual and hasty approach of students.

Preventing Calculation Errors

(i) Slow Down a Bit

Again, only slowing down and working more carefully on a problem will cut down on the computational errors.

(ii) Re-check the Calculation

After working hard to complete monotonous computations of multiple steps, students are usually reluctant to go back and check their calculations. However, testing the calculation for accuracy ensures whether the estimate is correct or not. If the last estimate is wrong, students should go back to their work and check for computational errors.

In my book, **"Be a Human Calculator,"** I have explained a unique method of checking calculations (Addition, subtraction, multiplication, division, squaring, cubing, square root, cube root, etc.) without doing the calculation again. The method is known as 'Digit Sum Method' and students can learn it quickly after some practice.

(iii) Improve Calculation Skills

Students should try to improve their calculation speed by using some good book. The time saved by them due to faster calculation skills may be utilized to complete the test early and this time may further be used to go through the answer-sheet once again to check elimination of possible silly mistakes. However, this advice will only apply to countries like India where calculators are not allowed in schools till Class XII.

For example, my book **"Be a Human Calculator"** provides lots of pure observation based tricks and hands-on practice questions for improvement of calculation skills of students in Arithmetic and Algebra.

3. Conceptual or Basic Silly Mistakes

Conceptual errors occur because students have misunderstood the underlying concepts or have used false logic. It is the most severe fault to find, but it is the most important to catch and correct.

When students make these types of mistakes, it's possible that all their math calculations are correct. If they've misunderstood a concept and thus used an incorrect method to solve, there is a possibility that they can work out each step meticulously and correctly but still get the wrong answer.

Preventing Conceptual Errors:

Apparently, avoiding conceptual errors is not as easy or straightforward as careless or computational errors. And of course, all students will have varying degrees of understanding and will struggle with different concepts.

But here are a few things students can try to encourage conceptual understanding and prevent future conceptual mistakes.

(i) Understand the Why

Students are advised to explore and discover new math concepts in a way that helps them to see and understand the why behind every idea of mathematics. Merely learning formulae and steps of the solution will not suffice. It is not always easy, but knowing the why behind math properties or methods will help students to understand, form connections and keep the information for a longer duration.

(ii) Multi-Channel Approach

There is usually more than one approach to solve a math problem. By learning or exploring a concept in multiple ways and from various angles, students may give themselves a richer math environment and deeper understanding

For attaining this objective, students are advised to learn various concepts by using a variety of sources which may be books, teachers, friends, online resources, etc.

(iii) Ask Your Teacher

Students are usually hesitant in asking doubts from teachers. It may happen due to their shy nature or due to the inhibitions that their doubts may seem very silly to other students and educators.

Students should note that to learn something new they have to shed such inhibition and therefore, students are advised to discuss all their doubts with the teachers. It will show students' understandings as well as misconceptions to teachers. That will further help the teachers to understand the level of knowledge of students, and they will be able to help students accordingly.

These doubts can also allow students to explain things in their words which may inspire other students as well to ask their doubts.

(iv) Analyzing Errors

In the early school years, it can be difficult for the students to analyze errors, because so much of what they are learning is computational. It can be easy at that stage to dismiss mistakes as merely calculation mistakes without worrying about other aspects of mathematics.

By middle or high school, students can recognize and classify their mistakes themselves. They should be allowed to think through what types of errors they're making so that they can, in fact, learn from them.

Apart from this chapter, this book will focus on the conceptual aspect of silly mistakes only. I have tried my best to compile 100 such stupid errors/misconceptions held by the students across the globe over the years, gathered from my teaching experience of 25 years.

So, let us start our journey to explore the world of silly mistakes to make a full stop on future stupid mistakes.

2
Be Sure About Division

i. Division of 0 by any finite number

Zero is the mathematical representation of the concept of "Nothing" and the operator of division on the other hand, is a mathematical tool to represent the process of “dividing or splitting some objects or numbers into smaller parts or pieces.”

For learning this simple concept, first, we have to revisit the simple idea of division. If a group of 4 persons has 20 bananas among them, then, each would get 20/4 = 5bananas. Thus, division means sharing or breaking or splitting some objects equally among some persons. Extending the concept, if on a given day these four persons have 0 bananas, then, the share of each on that day will be 0,i.e., 0/4 = 0.

We may extend the same concept to divide "Nothing (0)" among any number of finite people. So zero divided by any finite number is same as "Sharing nothing among any number of finite people." Each person gets nothing, i.e., 0/N = 0 for any finite number N except 0.

ii. Meaning of x/0

Explanation -1:

Let x is a non-zero finite number,

And also, let x/0 is defined and is equal to y (x / 0 = y).

Then, x = 0, but x is a non-zero finite number.

Therefore, x/0 is not defined.

Explanation -2:

If we divide six objects amongst two persons,

Then, share of each person = 6/2 = 3

But, if we divide, six objects amongst 0 persons,

The share of each individual=6 / 0 = not defined, as there is no person to claim the share.

From the above explanations, we may see that any finite number divided by 0 is not defined.

iii. Meaning of 0/0

Students should note that x/x =1 is valid only when x is not equal to zero.

Explanation-1

We know that,

If a/b = c, then b x c = a

On the same analogy, we may say that:

0/0 = 1 is true as 0 x 1 = 0

0/0 = 2 is true as 0 x 2 = 0

0/0 = 3 is true as 0 x 3 = 0 and so on.

Therefore, we may see from above that 0/0 may assume any value and hence it is not defined.

Explanation -2:

If we divide six objects amongst two persons,

Then, share of each person = 6/2 = 3

But, if we divide, 0 objects to 0 persons,

The share of each = 0 / 0 = not defined, as there is no person to claim the share and no objects for distribution.

iv. Solution of a ÷ b ÷ c

The concept of second division is confusing when there is no bracket.

So, (24 ÷ 4) ÷ 2 = 6 ÷ 2 =3

And, 24 ÷ (4 ÷ 2) = 24 ÷ 2 =12

But what about - 24 ÷ 4 ÷ 2,

Is it 3 or 12? Confused!!!

When more than one division symbols are involved in a question with no brackets, the golden rule is **first-cum-first.**

Thus, 24 ÷ 4 ÷ 2 = 6 ÷ 2 =3 and similarly,

100÷ 10 ÷ 2 ÷ 5 = 10 ÷ 2 ÷ 5 = 5 ÷ 5 =

3

Be Comfortable With Numbers

i. Wrong Definition of Prime Numbers

We can recall from our school days that number divisible by 1 and itself only are called prime numbers, but it is a wrong definition. Also, the definition that one is neither prime nor composite is partly wrong.

If we apply the above definition on 1 (one), then it should be a prime number as it is divisible by 1 and itself.

We may correct the above definition as follows:

Numbers (other than 1) divisible by 1 and itself are prime numbers.

However, the proper definition of prime and composite numbers is as follows:

The numbers having exactly two factors are prime numbers. Ex: 2, 3, 5, 7 etc. All these numbers have exactly two factors, 1 and number itself.

The numbers having more than two factors are composite. Ex : 4 (factors 1,2,4) , 6 (F 1,2,3,6) etc.

Now, 1 is a special number which does not fall under any of these two categories (Prime and Composite) as it is the only number with exactly one factor and it is neither prime nor composite.

ii. Are -3, -5, -7 etc. prime numbers?

Prime Numbers and Composite Numbers come from the class of Natural Numbers (Counting Numbers), and therefore negative numbers do not fall under the category of Prime or Composite Numbers.

iii. Are -3,-5,-7, etc. odd numbers?

All integers, irrespective of being positive or negative, divisible by 2 are even.

Even and odd numbers come under the category of integers. The integers which are wholly divisible by 2 are even integers and the integers, which are not wholly divided by 2, are odd integers.

Hence, -3,-5,-7,etc. odd numbers

iv. Are -2,-4,-6,etc. even numbers?

-2,-4,-6, etc.are even numbers as these numbers are integers and divisible by 2.

v. Are 1.2, 2.8, 3.4,etc. even numbers?

Even and odd numbers come from the class of integers. Decimal numbers are not integers. Therefore, decimal numbers are neither even nor odd.

vi. Can multiples of a number be negative?

Typically, like factors, multiples of a number are also natural numbers or positive integers. For example, multiples of 6 are 6, 12, 18 and many more. That means if otherwise specified, multiples of a number means natural numbers divisible by the given number.

However, in rare conditions or methods, we use negative multiples as well.

For example in question, Sum of first three negative multiples of 5 is x, find x. We will take first three negative multiples of 5 as -5, -10 and -15 and so the answer will be -30.

vii. Can factors of a number be negative?

Usually, factors of a number are natural numbers or positive integers. For example, factors of 12 are 1, 2,3,4,6 and 12. That means, typically, factors of a number mean natural numbers which divide the given number.

However, in rare conditions or methods, we use negative factors as well. For example, while factorizing a polynomial of higher degree, we use all the factors (negative and positive) of constant term to factorize that polynomial.

For example,

While factorizing $x^2 - 7x + 12$,

we factorize 12 as (-4 x -3) to split the middle term as,

$x^2 - 4x - 3x + 12 = x(x-4) - 3(x-4) = (x-4) . (x-3)$

viii. The Identity of a numbers

The identity of a real number is a number which on operation with the given number returns you the original number.

a + 0 = a = 0 + a (Additive Identity)

a x 1 = a = 1 x a (Multiplicative Identity)

We may see that, when we add 0to any real number, it gives us back the same number and hence 0 is known as the additive identity.

On the other hand, when we multiply 1with any real number, it gives us back the same number, and hence 1 (one) is known as the multiplicative identity.

Thus, the identity of a real number depends upon the operation we use.

There is no identity for subtraction and division.

It looks that 0 is the identity for subtraction also, but it is not as:

$a - 0 = a$, but $0 - a$ is not equal to a.

Also, it looks that 1 is the identity of division too, but it is not the case as:

$a / 1 = a$ but $1/a$ is not equal to a.

ix. The inverse of a number

We may define the inverse of a real number as under:

For addition

Number + Additive Inverse = Additive Identity (0)

Therefore,

Additive Inverse = - Number

For Multiplication

Number x Multiplicative Inverse = Multiplicative Identity (1)

Therefore,

Multiplicative Inverse = 1/ Number

Therefore, for any real number a,

Multiplicative inverse is 1/a;

And the additive inverse is -a.

x. Confusion over prime and co-prime

Students are always confused about the terms prime numbers and co-prime numbers. Let us clear the doubt by learning the correct definition of the two terms:

Prime Numbers- Numbers having exactly two factors or divisible by exactly two numbers are prime numbers. For example, 2,3,5,7, etc. all are prime numbers because these numbers are divisible by exactly two numbers, i.e., 1 and the number itself.

Co-prime Numbers- Two numbers are coprime to each other if they have no other factor as common other than 1 or their common factor is one only. For example, 2 and 3 are co-prime to each other as their common factor is 1. Similarly, 7 and 8 (not prime) are co-prime to each other as they have their common factor as 1.

xi. π is equal to 22/; still, it is irrational

We know that any number of the form p/q, where $q \neq 0$, is called rational number.

Now, a question arises, if π is equal to 22/7, why is it termed as irrational?

The answer lies in the fact that π is a non-terminating and non-repeating decimal and its value is 3.14159.........

Also, all non-terminating and non-repeating decimals are irrational as these numbers are not of the form of p/q.

22/7 is only the closest rational approximation to π for making calculations involving π a little easier.

xii. Why non-terminating and repeating decimals are rational numbers

See the following examples:

Example-1

Let x = .333……

10 x = 3.33…….

On subtraction, we get

9x = 3 or x = 3/9 = 1/3

Example-2

Let x = .353535……

100 x = 35.3535…….

On subtraction, we get

99 x = 35 or x = 35/99

From these examples, it is evident that we may convert every non-terminating and repeating decimal in the form p/q (where q ≠ 0 and p and q are integers) and thus these numbers are rational numbers.

4
Be More Positive About Zero

i. 0 is a Prime or Composite

Prime Numbers and Composite Numbers is a classification for Natural Numbers (Counting Numbers), and therefore zero does not fall under the class of Prime or Composite Numbers.

ii. 0 is even or odd

All integers, irrespective of being positive or negative, divisible by 2 are even. Because 0 is divisible by 2 and is an integer, it is an even number.

iii. Is Zero a Multiple of Every Number?

As explained in the earlier topic, usually (if mentioned otherwise), Multiples of a number are natural numbers only, and so zero does not form part of multiples of any number.

However, there are situations/methods where we talk about an integral multiple of a number.

For example, if x is an integral multiple of y,

we have x = I * y, where I is an integer.

For I=0, we may have x = 0 for every value of y.

And so, in this case, we can say that 0 is a multiple of every number.

iv. Multiples of 0

When we multiply 0 by any number, we get the result as zero.Therefore, 0 is the only whole number having only one multiple, i.e., 0 itself

All the other numbers have an infinite number of Multiples.

v. Factors of 0

Factors are the numbers which divide the given number. For example factors of six, are 1, 2, 3 and 6 as six are divisible by these numbers.

0 is divisible by every integer (0/1 =0, 0/2=0, 0/3=0, 0/-2=0 etc.) so all the integers except 0 are factors of zero.

Every natural number or positive integer has a finite number of factors.

Zero is the only number which has infinite factors.

vi. Inverse of 0

From the discussion held on the last page,

We have,

Multiplicative inverse and additive inverse of a number a as $1/a$ and $-a$ respectively.

Thus,

Additive Inverse of 0 = - 0 = 0

Multiplicative Inverse of 0 = 1/0 i.e. not defined.

5
Make Exponents Your Friend

i. Why $x^m . x^n = x^{m+n}$?

We know that $x. x = x^2$

Similarly, $4 \times 4 \times 4 = 4^3$

Thus, writing an index over a number or a term shows that how many times we have multiplied that number or term.

Thus,

$x^m . x^n$

$(x.x.x.x\ldots\ldots m \text{ times}).(x.x.x.x\ldots\ldots n \text{ times})$

$[x.x.x.x\ldots\ldots(m+n) \text{ times}]$

x^{m+n}

For example:

$2^7 . 2^{12}$

$(2.2.2.2\ldots\ldots 7 \text{ times}).(2.2.2.2\ldots\ldots 12 \text{ times})$

$[2.2.2.2\ldots\ldots 19 \text{ times}] = 2^{19}$

ii. Why $x^m / x^n = x^{m-n}$?

x^m / x^n

$(x.x.x.x$……..m times$)$ / $(x.x.x.x$……..n times$)$

$[x.x.x.x$…….. (m-n) times$]$ (due to cancellation)

x^{m-n}

For example:

$2^{12}/2^7$

(2.2.2.2……..12 times) /(2.2.2.2……..7 times)

[2.2.2.2……..5 times]

2^5

(As out of Twelve 2's in the numerator, Seven 2's have been cancelled out from seven 2's in the denominator leaving five two's in the Numerator).

iii. Why $(x^m)^n = x^{m.n}$?

$(x^m)^n$

$(x^m).(x^m).(x^m).(x^m)$…………n times

$x^{m+m+m+m \ldots \ldots \text{n times}}$

$x^{m.n}$

For example:

$(2^{12})^3$

$2^{12} . 2^{12} . 2^{12}$

$2^{12+12+12}$

$2^{12x\ 3}$

2^{36}

iv. Why $x^0 = 1$ (where $x \neq 0$)?

We know that, $x/x = 1$ **(where $x \neq 0$)**----- (1)

Also, $x^m/x^n = x^{m-n}$**(law of exponent)** ------- (2)

From (2) above, we may have

$x/x = x^{1-1} = x^0$ ------ (3)

From (1) and (3), we have $x^0 = 1$

v. If x^0 is 1, why $0^0 \neq 1$?

We know that, $x/x = 1$ ------ (1)

(Where $x \neq 0$)

We know that $x/x = x^{1-1} = x^0$------ (2)

From (1) and (2), we have $x^0 = 1$

But for $x = 0$, $x^0 = x/x \neq 1$

Hence,

$0^0 \neq 1$

vi. Why $x^{-m} = 1/x^m$?

we know that,

$x^0 = 1$ ----(1)

and

$x^m / x^n = x^{m-n}$ ----(2)

From (1) and (2), we have

$1/x^m = x^0 / x^m = x^{0-m} = x^{-m}$

vii. Why $x^m . y^m = (x.y)^m$?

We know that,

$x^m . y^m$

$(x.x.x.x\ldots\ldots$m times$).$ $(y.y.y.y\ldots\ldots$m times$)$

We may re-group these terms as follows,

$[(x.y). (x.y). (x.y) \ldots\ldots\ldots \text{m times}]$

$(x.y)^{m}$

For example:

$2^5.3^5 = (2.3)^5 = 6^5$ is true,

$3^7.5^7 = (3.5)^7 = 15^7$ is true,

But,

$10^5.3^3 = (10.3)^{5+3} = 30^8$

and

$10. (2.5)^2 = (25)^2 = 625$ are wrong.

viii. Silly Mistakes while using exponents

Some common silly mistakes committed by students while attempting the questions on exponents have been given below. These mistakes may easily be improved by observing basic rules of exponents explained in previous pages:

Mistake-1: $x^2 + x^3 = x^5$

Correct Solution: $x^2 + x^3 = x^2 (x + 1)$

Mistake-2: $x^2 . x^3 = x^6$

Correct Solution: $x^2 . x^3 = x^5$ **(Use $x^m . x^n = x^{m+n}$)**

Mistake-3: $x^6 - x^2 = x^4$

Correct Solution: $x^6 - x^2 = x^2 (x^4 - 1)$

Mistake-4: $x^6 / x^3 = x^2$

Correct Solution: $x^6 / x^3 = x^3$ **(Use $x^m / x^n = x^{m-n}$)**

Mistake-5: $(x^2)^3 = x^5$

Correct Solution: $(x^2)^3 = x^6$ [Use $(x^m)^n = x^{m.n}$]

Mistake-6: $10. (2)^3 = 20^3$

Correct Solution: $10. 8 = 80$

6
Silly Mistakes in Arithmetic/Algebra

i. Sign Convention while adding

The careless mistake of using a wrong sign is not only common during early school years, but students continued these errors even at higher level.

There are two golden rules of applying sign while adding:

(a) Add the numbers with same sign and subtract the numbers with opposite sign;

(b) Answer has the sign of number with greater numerical value;

Thus, $5+4=9$

$(-5)+(-4)=-9$

$(5)+(-4)=1$

$(-5)+(4)=-1$

ii. Sign Convention while subtracting

Apart from the two rules followed for addition, students should follow one more rule for subtracting numbers:

"While subtracting, change the sign of the second number."

Thus, 5- 4 = 1

5 - (-4) = 5 + 4 = 9

(-5) - (-4) = -5 + 4 = -1

-5 - 4 = -9

iii.Sign Convention while multiplying/ dividing

For multiplication/divisionof numbers, the sign convention is governed by the following rule:

"When the sign of two numbers is identical, the answer is positive, and when the sign of two numbers is opposite, the answer is negative."

Thus,

(+6).(+2) = +12 and(+6) / (+2) = + 3

(-6). (- 2) = +12 and(-6) / (-2) = +3

(+6). (-2) = -12 and (+6) / (-3) = -2

(- 6).(+2) = -12 and(-6) / (+2) = - 3

iv. Comparing Negative Numbers

In negative numbers, the number with less numerical value is greater.

Mistake-1: $-7 > -2$

Correct Solution: $-7 < -2$

Mistake-2: $-5 < -9$

Correct Solution: $-5 > -9$

v. Comparing Fractions

When Numerators of the numbers are same, the number with lesser denominator is greater.

Mistake-1: $1/7 > 1/5$

Correct Solution: $1/7 < 1/5$

Mistake-2: $3/5 < 3/8$

Correct Solution: $3/5 > 3/8$

vi. Finding LCM of Numbers

We know that in Algebra, Least Common Multiple (LCM) of two numbers a and b is $a.b$.

Students try to imitate the same in arithmetic and end up having a wrong answer.

Thus,

LCM of 2 and 3 is 6;

LCM of 3 and 4 is 12;

But LCM of 4 and 6 is not 24; it is 12.

Thus, LCM of a and b is $a.b$ is true in arithmetic only when both the numbers are co-prime to each other. In other cases, students have to find the answer by usual methods.

vii. Every real number has two square roots

It is a widespread misconception among students, and it is known as **"Square Root Fallacy."**

In fact, every real number has only one square root by default, and that is positive square root, thus:

$$\sqrt{9} = 3, \sqrt{25} = 5, \sqrt{100} = 10$$

Let us assume, there are two square roots, one is positive and one is negative. Then, we have:

$3 = \sqrt{3 \times 3} = \sqrt{(-3) \times (-3)} = -3$, which is apparently wrong.

Now, a question arises that if every real number has only one square root that why we have two answers to questions like:

Find x, if $x^2 = 9$. Answer of this question is ± 3 **(why).**

The above question is different from finding the square root of 9 i.e. $\sqrt{9} = 3$.

Here, we have to find a number whose square is 9, and there may be two such numbers one is positive, and one is negative. Thus, the correct way of solving this will be as follows:

$$x^2 = 9 \Rightarrow x = \pm\sqrt{9} = \pm 3$$

Students should, therefore, note that

$$\sqrt{a^2} = a \text{ whereas}$$

$$x^2 = a^2 \text{gives } x = \pm\sqrt{a^2} = \pm a$$

viii. Square root of Product of Two Numbers

The above statement is wrong and leads to erroneous results. We may verify it by seeing the following example:

$$1 = \sqrt{1.1} = \sqrt{(-1).(-1)} = \sqrt{-1}.\sqrt{-1}$$

$$= i.i = i^2 = -1$$

What went wrong?

The formula, $\sqrt{a.b} = \sqrt{a}\,.\sqrt{b}$, can be used only when at least one out of a and b is positive. If both the numbers are negative, the formula will not work.

ix. Squaring a negative number

This mistake is usually carried out by students during early years of schooling, but sometimes it happens afterward as well due to oversight.

$(-3)^2 = -3^2 = -9$

With a little caution, we know that the answer is:

$(-3)^2 = -3 \times -3 = 9$

x. Careless cancellation

(a) Observe the careless cancellation in the following examples:

$\frac{2x+3}{2} = x + 3$ (Incorrect); $\frac{2x+3}{2} = x + \frac{3}{2}$(correct)

$\frac{7x+5}{5} = 7x + 1$ (Incorrect); $\frac{7x+5}{5} = \frac{7}{5}x + 1$(correct)

(b) Biggest mistake in algebraic cancellations:

Cancellation is nothing but a shortcut for division, thus 5x = 5 means x=5/5 = 1.

However, to save time we use cancellation as follows:

$7x = 49$ means $x = 7$

$12x = 48$ means $x = 4$

Everything is fine as long as we are dealing with numbers. Consider the following example:

$a^2 = a.b$ means $a = b$**(correct or incorrect).**

Wrong, as $a^2 = ab \Rightarrow \frac{a^2}{a} = b \Rightarrow a = b$, includes one wrong step as we are dividing by algebraic terms and the same will be valid if the denominator is not zero. Thus, $\frac{a^2}{a}$ will be equal to a, only when $a \neq 0$. If $a = 0$, the term $\frac{a^2}{a}$ will become 0/0 and will not be defined.

The correct way of solving such questions will be as under:

$a^2 = ab \Rightarrow a^2 - ab = 0 \Rightarrow a(a - b) = 0 \Rightarrow a = b$ or $a = 0$.

So, students should cancel out any algebraic term only when that algebraic term is non-zero.

Thus,

The solution of $x^2 = x$ ($where\ x \neq 0$) is x =1 (here cancellation by x is valid as it is non-zero).

However, the solution of $x^3 = x^2$ is $x = 1$ is invalid as nothing is given about x. Thus the correct solution will be $x^3 - x^2 = 0$ or $x^2 (x-1) = 0$ or $x = 0$.

xi. Casual use of brackets

(a) Incomplete brackets never give a clear picture.

Hence, in the expression, $5x^3 - (7x^2 - 8x - 1$, It is not clear whether the intention was to write-

$5x^3 - (7x^2 - 8x - 1) \qquad = 5x^3 - 7x^2 + 8x + 1$

Or $5x^3 - (7x^2 - 8x) - 1 \qquad = 5x^3 - 7x^2 + 8x - 1$

Both the answers are different. Thus, an incomplete bracket may sometimes leave students in difficulty.

(b) For multiplying a - b and $a + b$, if we write these numbers without using brackets as $a - b.\ a +$

b, we will end up getting the answer as $a - a.b + b$, which is wrong.

The correct steps are,

$(a - b).(a + b) = a^2 - b^2$

Thus, whenever, we multiply algebraic terms of order two or more, it is advisable to use brackets.

(c) For subtracting a - b and $a + b$, if we write these numbers without using brackets as $a - b - a + b$, we will end up getting the answer as 0, which is wrong.

The correct steps are,

$(a - b) - (a + b) = a - b - a - b = -2b$

Thus, whenever, we subtract algebraic terms of order two or more, it is advisable to use brackets.

(d) One more example of careless use of brackets is

$2 \int (5x^4+7)\, dx = 2x^5+7x+C$

The correct solution should be:

$2\int(5x^4+7)\,dx = 2(x^5+7x)+C = 2x^5+14x+C$

xii. Using Double Sign

Using two signs together is not acceptable as done in the following examples:

Add x and – 5

Wrong Way: $x + -5$

Right Way: $x + (-5) = x - 5$

Multiply x and – 5

Wrong Way: $x. -5$

Right Way: $x. (-5) = -5x$

xiii. Wrong Notations

(a) Using one as a coefficient or as an exponent:

$1x$and x^1 are two frequently used examples of wrong notations. As per mathematical convention, when the coefficient or the power of a term is 1, it is better to write it simply as x in place of $1x$and x^1 respectively.

(b) Using / to denote a fractional coefficient: Second example of bad notation is using a / sign to denote a fractional coefficient before a variable: Suppose we want to write $\frac{4}{5}x$, and if we write it as 4/5x, it is not clear whether it is $\frac{4}{5}x$ or $\frac{4}{5x}$ and thus it is ambiguous in nature.

(c) Using / to denote a fraction in Algebra: The third example of bad notation is using a slash ('/ ') sign to represent a fraction involving algebraic terms:

Suppose we want to write $\frac{x+y}{x-y}$ and if we write it as $x + y/x - y$, it will not be clear whether it is $\frac{x+y}{x-y}$or $x + \frac{y}{x} - y$ and thus ambiguous in nature.

However, if students wanted to use a slash ('/ ') as a substitute for the division, they should use brackets to write the above expression as $(x + y)$ / $(x - y)$.

(d) Wrong notation for Square Root: We express cube root of a number x as $\sqrt[3]{x}$, the fourth root as $\sqrt[4]{x}$. Keeping the same analogy in mind, sometimes student used to write the square root of a number as $\sqrt[2]{x}$, which is not acceptable. We should write the square root only as $\sqrt{x}$.

(e) Putting Equal Sign before two equivalent Equations:

Just see the following calculations-

$7x + 3 = 2x + 8$

$= 7x - 2x = 8 - 3$

$= 5x = 5$

$= x = 1$

It is evident from the above example that an equal sign is used before two equivalent equations incorrectly. The correct way of calculation is as under:

$7x + 3 = 2x + 8$

$7x - 2x = 8 - 3$

$5x = 5$

$x = 1$

(f) Using coefficient between two variables:

Avoid expressions like $x2y$, $a3b$, etc

Write the above expressions as $2xy$and $3ab$

In algebra, the convention is to write constants terms before the variable.

xiv. Improper Distribution

Students are aware of distribution law which works as under:

$a.\ (b \pm c) = a.b \pm a.c$ (Distribution of Multiplication over Addition or Subtraction)

However, students take the liberty to extend the law wrongly as,

$\sqrt{x+y} = \sqrt{x} + \sqrt{y}$ (Square Root is mistakenly distributed over Addition)

$\frac{a}{x+y} = \frac{a}{x} + \frac{a}{y}$(Division is wrongly distributed over Addition)

Both of the above expressions are wrong as will be evident from following examples:

$\sqrt{5+4} \neq \sqrt{5} + \sqrt{4}$

$\frac{6}{2+1} \neq \frac{6}{2} + \frac{6}{1}$

xv. Cancellation in Inequalities

See the following inequalities:

$2x > 6 \Rightarrow x > 3$ (Divide both sides by 2)

$\frac{x}{4} < 3 \Rightarrow x < 12$(Multiply both sides by 4)

$-2x > 6 \Rightarrow x > -3$ (Divide both sides by -2)

$-3\,x < -12 \Rightarrow x < 4$ (Divide both sides by -3)

$\frac{x}{-4} < 3 \Rightarrow x < -12$ (Multiply both sides by-4)

Out of the above examples, first two examples are right, whereas last three examples are wrong.

Whenever both sides of the inequality are multiplied or divided by a negative number, the sign of the inequality gets reversed.

Thus, the correct solution of last three examples will be as under:

$-2x > 6 \Rightarrow x < -3$ (Divide both sides by -2)

$-3x < -12 \Rightarrow x > 4$ (Divide both sides by -3)

$\frac{x}{-4} < 3 \Rightarrow x > -12$ (Multiply both sides by-4)

xvi. Quadratic Inequalities

A prevalent mistake, I have observed over the years is as follows:

It is true that, $x^2 = 16 \Rightarrow x = \pm 4$,

On the above lines, students are tempted to write,

$x^2 > 16 \Rightarrow x > \pm 4$, and

$x^2 < 16 \Rightarrow x < \pm 4$, which are wrong.

So, students may take a note that,

$x^2 = a^2 \Rightarrow x = \pm a$, is true only for an equation.

The correct solutions of inequalities of these forms are as under:

$x^2 < a^2 \Rightarrow -a < x < a$

$x^2 \leq a^2 \Rightarrow -a \leq x \leq a$

$x^2 > a^2 \Rightarrow -a > x > a$

$x^2 \geq a^2 \Rightarrow -a \geq x \geq a$

xvii. Inequalities Involving Modulus

Just like the silly mistake committed by students in quadratic inequalities, a similar mistake is committed by them while solving inequalities involving modulus of number:

On the lines of $|x| = 4 \Rightarrow x = \pm 4$,

Students are tempted to write,

$|x| > 4 \Rightarrow x > \pm 4$

And $|x| < 4 \Rightarrow x < \pm 4$, which is wrong.

So, students may take a note that,

$|x| = a \Rightarrow x = \pm a$, is true only for an equation.

The correct solutions of inequalities of these forms are as under:

$|x| < a \Rightarrow -a < x < a$

$|x| \leq a \Rightarrow -a \leq x \leq a$

$|x| > a \Rightarrow -a > x > a$

$|x| \geq a \Rightarrow -a \geq x \geq a$

xviii. Wrong use of algebraic identities

Due to carelessness, many algebraic identities are used incorrectly by the students. Some of these are:

$(a \pm b)^2 = a^2 \pm b^2$

$(a \pm b)^3 = a^3 \pm b^3$

$a^2 - b^2 = (a - b)^2$

$a^3 \pm b^3 = (a \pm b)^3$

$a^3 + b^3 + c^3 = 3abc$

$(-a - b)^2 = -(a + b)^2$

The correct identities are:

$(a \pm b)^2 = a^2 + b^2 \pm 2ab$

$(a \pm b)^3 = a^3 \pm b^3 \pm 3ab\,(a \pm b)$

$a^2 - b^2 = (a - b)\,.\,(a + b)$

$a^3 \pm b^3 = (a \pm b).(a^2 + b^2 \mp ab)$

$a^3 + b^3 + c^3 = 3abc$only when $a + b + c = 0$

$(-a - b)^2 = (a + b)^2$

xix. Real numbers are not polynomials

A polynomial is an algebraic expression consisting of products of constants and exponents of variables separated by addition or subtraction where exponents are non-negative integers.

A polynomial in a single variable x can always be written (or rewritten) in the form:

$a_n x^n + a_{n-1} x^{n-1} + a_{n-2} x^{n-2} + \ldots\ldots\ldots\ldots + a_2 x^2 + a_1 x + a_0$

Here all the exponents are non-negative integers or whole numbers and all the coefficient $a_0, a_1, a_2 \ldots\ldots a_n$ are real numbers.

So any algebraic expression satisfying the above criteria is a polynomial.

Now, consider a real number 5.We may express it as $5.x^0$. It is a polynomial as coefficient five is a real number, and exponent 0 is a whole number.

So, we may express any real number r in the form of $r.x^{0}$, and so it is a polynomial.

xx. Confusion over degree and order

Students are always confused about these two terms about the degree and order of a polynomial.

Order: The number of terms in a polynomial is called its order.

Thus, orders of $x + 2$, $x^2 + 4x + 7$ and -5 are 2, 3 and 1 respectively as the numbers of terms in these polynomials are 2, 3 and 1 respectively.

Degree: For a single variable polynomial, the highest power of the variable term used in that polynomial is called its degree.

For example, degrees of $x + 2$, $x^2 + 4x + 7$ and -5 are 1, 2 and 0 respectively as highest power of x in these polynomials is 1, 2 and 0 respectively.

For a multivariable polynomial, the largest sum of powers of the multi-variables in a term is called its degree.

Thus, for $x^2y + x^3y^2 + x.y + 5$, the degree is 5.

xxi. Leaving irrational denominator

It is a well-established convention in mathematics that while writing the answer in a fraction, the denominator should be a rational number.

In case, the denominator of a number is irrational; we make it rational by multiplying Numerator and denominator of the fraction by the conjugate of the denominator.

Consider the following answers:

$\frac{2}{\sqrt{3}}$ and $\frac{5}{\sqrt{3}-1}$.

In both the answers, denominators are an irrational number, which is against the mathematical convention.

So, to rationalize the denominator, we need to multiply $\sqrt{3}$ and $\sqrt{3}-1$ in the Numerator and Denominator of both the fractions respectively to get the correct answer as:

$\frac{2\sqrt{3}}{3}$ and $\frac{5(\sqrt{3}+1)}{2}$

7
Be Careful With Logarithm

i. What is Logarithm?

We may express any number as the power or index of some other number,

For example, $8 = 2^3$

(Here 2 is the base and 3 is the exponent)

We may write the above numbers as follows:

$\log_2 8 = 3$ (read as log eight to the base two is 3).

Thus, the logarithm is another way to express numbers written in exponent form. It has numerous applications in the field of mathematics, physics, and chemistry.

Thus, $x = a^y$ and $\log_a x = y$ convey the same meaning. However, students may note that there are some constraints while applying the concept of the logarithm. These constraints are as follows:

In $\log_a x$, $a \neq 1$, $a > 0$ and $x > 0$.

ii. Properties of Logarithm

Just like exponents, the concept of logarithm also follows certain properties mentioned below:

(a) $\log_a (x.y) = \log_a x + \log_a y$

(b) $\log_a (x/y) = \log_a x - \log_a y$

(c) $\log_a (x)^m = m.\log_a x$

(d) $\log_a a = 1$

(e) $\log_a 1 = 0$

(f) $\log_a b = 1/\log_b a$

(g) $\log_c b = \log_a b / \log_a c$

(h) $b^{\log_b x} = x$

Most of the silly mistakes committed while using logarithm may easily be avoided by having a close look at these properties.

iii. Using powers in Logarithm

In trigonometry,

We write $\sin x. \sin x$ as $(\sin x)^2$ or $\sin^2 x$

However,

In logarithm,

It is wrong to use $\log^2_a x$ for $(\log_a x).(\log_a x)$.

The right way of expressing $(\log_a x).(\log_a x)$is $(\log_a x)^2$ only.

iv.Confusion Over log x and ln x

Students are always confused about the symbols $\log x$ and $\ln x$.

Log x is the short form of $\log_{10} x$ (when the base is 10). We call log x or $\log_{10} x$ as the **common logarithm.**

ln x is the short form of $\log_e x$ (when the base is e). We call $\ln x$ or $\log_e x$ as the **natural logarithm**. Students will know more about the number ' e ' in chapter 9.

v. Silly Mistakes while using Logarithm

See the following mistakes frequently committed by students:

Mistake-1: $\log_a (x/y) = \log_a (x - y)$

Correct Solution: $\log_a (x/y) = \log_a x - \log_a y$

Mistake-2: $\log_a x - \log_a y = (\log_a x) / (\log_a y)$

Correct Solution: $\log_a x - \log_a y = \log_a (x/y)$

Mistake-3: $\log_a (b.x^n) = b.\log_a(x^n)$

Correct Solution: No such property exists. The correct solution will be as under:

$\log_a(b.x^n) = \log_a b + \log_a x^n = \log_a b + n.\log_a x$

Mistake-4: $\log_a (x.y) = (\log_a x). (\log_a y)$

Correct Solution: The property is used incorrectly. The correct solution will be as under:

$\log_a(x.y) = \log_a x + \log_a y$

Mistake-5: $\log_a (x/y) = (\log_a x) / (\log_a y)$

Correct Solution: The property is used incorrectly. The correct solution will be as under:

$\log_a(x/y) = \log_a x - \log_a y$

Mistake-6: $\log_a x = \log ax$

Correct Solution: In $\log_a x$, the base is a, whereas in $\log ax$, the base is not written and hence it is the short form of $\log_{10} ax$.

Mistake-7: $\log_a x^n = (\log_a x)^n$

Correct Solution: The correct property is:

$\log_a (x)^n = n.\log_a x$

Mistake-8: $\log_a (x + y) = \log_a x + \log_a y$

Correct Solution: The correct property is:

$\log_a(x.y) = \log_a x + \log_a y$

Mistake-9: $\log_a (x - y) = \log_a x - \log_a y$

Correct Solution: The correct property is:

$\log_a(x/y) = \log_a x - \log_a y$

Mistake-10: $b^{-\log_b x} = -x$

Correct Solution: The correct solution is:

$b^{-\log_b x} = b^{\log_b x^{-1}} = b^{\log_b (\frac{1}{x})} = 1 / x$

BY THE SAME AUTHOR

8

Silly Mistakes in Trigonometry

i. Sin A = Sin. A

In trigonometry, students are often confused while writing trigonometric ratios and consider Sin A as the product of Sin and A, which is not true.

Sin A is a single ratio between opposite and hypotenuse when the base angle (the angle at one vertex of the base other than the right angle) is equal to A.

We know that, Sin 30^0= ½

Thus, in any triangle with the base angle as 30^0, the ratio of opposite and hypotenuse will always be ½.

We will see that, how this single omission on the part of students leads to other confusions in trigonometry.

ii. Cos (x + y) = Cos x + Cosy

This mistake is a by-product of the first mistake as we are treating *Cos (x + y) as Cos.(x+y)* and therefore, applying distribution law, which is wrong.

The correct identity for *cos (x + y)* is,

Cos (x + y) = Cos x. Cos y – Sin x. Sin y

iii. Sin nx =n.Sin x

Again as we are treating Sin nx as a product of sin.n.*x*, we are writing it as n. Sin x which is wrong.

Thus writing Sin 2*x* = 2.Sin *x* and Sin 3*x*= 3.Sinx is absolutely wrong.

In fact, we have identities for these expressions, and these are:

Sin 2*x*= 2.Sin *x*. Cos *x* and

Sin 3*x*= 3.Sin *x* – 4 $\sin^3 x$

iv. (Sin x) ^{2} *= Sin* x^2

In trigonometry,

sin x. sinx may be written as $(\sin x)^2$ or $\sin^2 x$

ButSin x^2 = Sin (x.x) $\neq(\sin x)^2$ or $\sin^2 x$

v. Cos^{-1}*x = 1/cos x*

In exponents, x^{-1}or reciprocal of x is written as $1/x$, which is perfectly fine. However, in trigonometry,

The reciprocal of cos x = 1/cos x, is correct,

$(\cos x)^{-1} = 1/\cos x$is also correct.

However, Writing$Cos^{-1}x = \frac{1}{Cos\ x}$ is not correct as $Cos^{-1}x$ or $(\cos x)^{-1}$ are not same as $\cos^2 x$ *or* $(\cos x)^{2.}$

The symbol $Cos^{-1}x$ is used to denote inverse trigonometric cosine function and not as $(\cos x)^{-1}$.

The correct way of expressing inverse trigonometric function is as under,

If y = $\sin x$

Then, x = $\sin^{-1} y$

vi. Confusion over degree and radian

Most of the students always try to solve a trigonometric question in degrees. It may be because of their habit until now, or it's easier for them to visualize the angles in degrees. But in practice, radian measure is used in most of the trigonometry and calculus, and students are often confused when it comes to using the radian over degrees.

Students are required to use radian measure while attempting questions of trigonometry or calculus if not mentioned otherwise.

We know that π radian = 180 degree or $\pi^c = 180^0$

For example,

$\sin(\pi/6)^c = \sin 30^0 = ½$ is correct,

But $\sin 30 = ½$ is incorrect

As in the second example, the symbol of the degree is missing, and students may assume it the degree measure, which is wrong.

In the case of absence of any symbol of degree or radian, students should understand that the angle is in radian.

vii. Length of the Arc of a circle

The angle subtended by an arc at the center of a circle (Θ), is given by the formula:

Θ = Length of arc (l) /Radius of the circle (r)

Now, see the following question:

Find Sin Θ, where Θ is the angle made by the arc of length 60 cm of a circle with radius 2 cm.

Solution:

$\Theta = l / r = 60 \text{ cm} / 2 \text{ cm} = 30$

Therefore, Sin $\Theta = \sin 30^0 = 1/2$

What went wrong?

The above formula works as under:

Θ (in radian) = Length of arc (l) /Radius of the circle (r)

Therefore, the correct solution is:

$\Theta = l / r = 60$ cm / 2 cm = 30 radian or 30^c

Therefore, Sin Θ = sin 30^c (which can be found using sin table).

9
Silly Mistakes in Calculus

i. Ignoring Notations for Limits

Students often forget to write the notations for limits while solving these questions after few steps as evident from the following example:

$$\lim_{x\to 3}\frac{x^2-9}{x-3}=\frac{(x-3)(x+3)}{x-3}=x+3=6$$

The correct way of writing this solution should be as under:

$$\lim_{x\to 3}\frac{x^2-9}{x-3}=\lim_{x\to 3}\frac{(x-3)(x+3)}{x-3}$$

$$=\lim_{x\to 3}(x+3)=6$$

Hence, students should keep writing the notation for limits up to the step in which they substitute the value of limit to the given function.

ii. Applying the limits on the part of a function

Students often use the value of limit on the part of function without simplifying the function altogether. It will be evident from the following example:

$$\lim_{x \to 0} \frac{ax + x\cos x}{b\,sinx} = \frac{1}{b}\lim_{x \to 0} \frac{x(a + \cos x)}{sinx}$$

$$= \frac{1}{b}\lim_{x \to 0} \frac{x(a + \cos 0)}{sinx} = \frac{1}{b}\lim_{x \to 0} \frac{x(a + 1)}{sinx}$$

$$= \frac{1}{b}\lim_{x \to 0} \frac{x}{sinx}.(a + 1) = \frac{a+1}{b}\lim_{x \to 0} \frac{x}{sinx} = \frac{a+1}{b}$$

$$[\because \lim_{x \to 0} \frac{x}{sinx} = 1]$$

In the above example, the limit is applied to some part of the Numerator without simplifying the entire function. The correct solution should be as under:

$$\lim_{x \to 0} \frac{ax + x\cos x}{b\,sinx} = \frac{1}{b}\lim_{x \to 0} \frac{x(a + \cos x)}{sinx}$$

$$= \frac{1}{b}\lim_{x \to 0} \frac{x(a + \cos x)}{sinx} = \frac{1}{b}\lim_{x \to 0} \frac{x}{sinx}.(a + cosx)$$

$$= \frac{1}{b}.1.(a + \cos 0) = \frac{a+1}{b}$$

iii. Improper use of L' Hospital's Rule

L' Hospital's Rule: As per this rule, for the limits of the following types:

$$\lim_{x \to a} \frac{f(x)}{g(x)} = \frac{0}{0} \ or \ \lim_{x \to a} \frac{f(x)}{g(x)} = \frac{\pm\infty}{\pm\infty}$$

Where a can be any real number, infinity or negative infinity. In these cases, we have,

$$\lim_{x \to a} \frac{f(x)}{g(x)} = \lim_{x \to a} \frac{f'(x)}{g'(x)}$$

Students have a tendency to use this formula as:

$$\lim_{x \to a}[f(x).g(x)] = \lim_{x \to a}[f'(x).g'(x)]$$

$$\lim_{x \to a}[f(x) \pm g(x)] = \lim_{x \to a}[f'(x) \pm g'(x)]$$

The above interpretation is wrong as L' Hospital's Rule is applicable only on the quotient of two functions and that too under certain conditions as specified above and not on product, sum or difference of two functions.

iv. Improper use of formula - derivative of x^n

$$\frac{d}{dx}x^n = nx^{n-1}$$

Students make a very common mistake in differential calculus by using the above result incorrectly. The students tend to forget that the above result is used only when the base number x is a variable and the exponent n is a constant (scalar) and not the vice versa and in any other similarly looking functions.

For example:

$$\frac{d}{dx}x^5 = 5x^{5-1} = 5x^4$$

is alright, but,

$$\frac{d}{dx}x^x = xx^{x-1} = x^x$$

is wrong as exponent x is a variable. The correct way of solving problems like this is as follows:

Let $y = x^x$,

$\log_e y = \log_e x^x$

$\log_e y = x.\log_e x$

$$\frac{d}{dx}\log_e y = \frac{d}{dx}x.\log_e x$$

$$\frac{1}{y}\frac{dy}{dx} = x\frac{d}{dx}\log_e x + \log_e x\frac{d}{dx}x$$

$$\frac{1}{y}\frac{dy}{dx} = x\frac{d}{dx}\log_e x + \log_e x.1$$

$$\frac{1}{y}\frac{dy}{dx} = 1 + \log_e x$$

$$\frac{d}{dx}x^x = x^x.(1 + \log_e x)$$

Similarly,

$$\frac{d}{dx}e^x = x.e^{x-1} \text{ and } \frac{d}{dx}a^x = x.a^{x-1},$$

are totally wrong as in these examples, the exponent x is a variable and the base numbers e and a are constants. The correct formulae for derivative of these functions are:

$\frac{d}{dx}e^x = e^x$

and

$\frac{d}{dx}a^x = a^x.\log_e a$

Therefore,

$\frac{d}{dx}5^x = 5^x.\log_e 5$

Important Note: When base is a variable and exponent is a constant use the formula

$\frac{d}{dx}x^n = n.x^{n-1}$

but when base is a constant and exponent is a variable use the formula

$\frac{d}{dx}a^x = x.a^{x-1}$

There is a separate formula for

$\frac{d}{dx}e^x = e^x$

where e is a constant but it is a special constant and so is separate from a as used in

$\frac{d}{dx}a^x$.

We will learn the difference between a^x and e^x in forthcoming topics.

v. Finding derivative of [f(x).g(x)] incorrectly

We know that,

$$\frac{d}{dx}[f(x) \pm g(x)] = \frac{d}{dx}f(x) \pm \frac{d}{dx}g(x)$$

but students often try to apply the same analogy to find derivative of product of two functions as :

or, $\frac{d}{dx}[f(x).g(x)] = \frac{d}{dx}f(x)\frac{d}{dx}g(x)$,

which is wrong.

The correct formula for finding derivative of the product of two functions is:

$$\frac{d}{dx}[f(x).g(x)] = g(x)\frac{d}{dx}f(x) + f(x)\frac{d}{dx}g(x)$$

vi. Finding derivative of [f(x)/g(x)] incorrectly

We know that,

$$\frac{d}{dx}[\mathrm{f(x)} \pm \mathrm{g\,(x)}] = \frac{d}{dx}\mathrm{f(x)} \pm \frac{d}{dx}\mathrm{g(x)}$$

But students often try to apply the same analogy to find derivative of the quotient of two functions as:

$$\frac{d}{dx}[\mathrm{f(x)}/\,\mathrm{g\,(x)}\,] = \frac{d}{dx}\mathrm{f(x)}/\frac{d}{dx}\mathrm{g(x)}$$

which is wrong.

The correct formula for finding derivative of the product of two functions is:

$$\frac{d}{dx}\left[\frac{\mathrm{f(x)}}{\mathrm{g(x)}}\right] = \left\{\mathrm{g(x)}\frac{d}{dx}\mathrm{f(x)} - \mathrm{f(x)}\frac{d}{dx}\mathrm{g(x)}\right\}/\{\mathrm{g(x)}\}^2$$

vii. Ignoring constants in Integration

We know that integration is the inverse process that of finding a derivative.

Thus if we have,

$\frac{d}{dx}$ (Sin x) = cos x,

We will also have ,∫cos x dx = sin x.

That means, if the derivative of sin x with respect to x is cos x, then, integration of cos x with respect to x will be sin x.

But,

$\frac{d}{dx}$ (Sin x) $=\frac{d}{dx}$ (sin x + 2) = $\frac{d}{dx}$ (sin x -10)…. and so on = cos x,

That gives,

∫cos x dx = sin x or sin x + 2or sin x -10, which shows that integration of a function may yield infinite values. So, it will be more appropriate to write, ∫cos x dx = sin x + c, where c is any arbitrary constant.

Due to ignorance or haste, students tend to drop the constant while finding the integration of some function which is not correct. Thus, ∫cos x dx = sin x is wrong, and students should write it as ∫cos x dx = sin x+ c as there are infinite values of integration of a function.

viii. Using double constants in Integration

We had seen in the last topic that while writing the answer for integration of a function, we need to add a constant to the solution to make the perfect sense. However, sometimes, students use two constants in the same question, which is not a good practice. For example: In the following question,

$$\int \frac{sin\ x}{\sin(x-a)}.dx$$

$$Let\ x - a = t\ \ or\ dx = dt$$

$$\int \frac{sin\ x}{\sin(x-a)}.dx = \int \frac{\sin(t+a)}{\sin t}.dt$$

$$= \int \frac{\sin t.cosa + cost.sina}{\sin t}.dt$$

$$= \int (\cos a + cost.sina).dt$$

$$= t.\cos a + sina.\log_e|sint| + C_1$$

$$= (x-a).\cos a + sina.\log_e|\sin(x-a)| + C_1$$

$$= x.\cos a + sina.\log_e|\sin(x-a)| - acos + C_1$$

Leaving the answer at this stage will be against the convention as there are two constants in the answer, -*a cos a* and C_1.

The correct way should be to write the resultant of -*a cos a* and C_1 as C and thus in the final solution, only one constant has been used. The answer should look like:

$$= x.\cos a + sina\,.\log_e|\sin(x-a)| + C$$

ix. Improper use of the formula for ∫xⁿ dx

Students often forget that there is a restriction on this integration formula, so the formula along with the restriction is under:

$$\int x^n\,.dx = \frac{x^{n+1}}{n+1},$$

provided $n \neq -1$

Thus it is wrong to use the formula where n = -1, in the following case:

$$\int \frac{1}{x}.\,\mathrm{dx} = \int x^{-1}\,.dx = \frac{x^{-1+1}}{-1+1} + c = \frac{x^0}{0} + c$$

The correct formula for finding the integration of 1/x is as follows:

$\int 1/\mathrm{x}.\,\mathrm{dx} = \log_e|x| + c$

x. Dropping the absolute value when integrating $\int 1/x\ dx$

We had seen in the previous topic that,

$\int 1/\mathrm{x}.\,\mathrm{dx} = \log_e|x| + c$

However, most of the students have a tendency to forget the sign of absolute value (modulus) attached with x in the formula, which is very much required.

Though, it is true that we do not require the notation for absolute value in some cases such as:

$$\int \frac{2x}{x^2+7}.\,dx = \log_e|x^2+7| + c$$

$$= \log_e(x^2+7) + c$$

In the previous example, the value of $x^2 + 7$ is positive. Therefore the use of absolute value notation has no meaning, and therefore its use is optional.

But consider the following case,

$$\int \frac{2x}{x^2 - 7} . dx = \log_e |x^2 - 7| + c$$

In this case, the value of x^2 - 7 may be positive or negative depending upon the value of x and so the use of absolute value notation is necessary.

Students are, therefore, advised to use the notation for absolute value in all cases wherever, the formula of

$\int 1/\text{x}.\,\text{dx} = \log_e |x| + c$ is used.

xi. Improper use of formula ∫1/x dx

We know that,

$$\int 1/\text{x}.\,\text{dx} = \log_e |x| + c$$

But this formula has been used by students in the wrong manner as will be evident from following examples:

$$\int 1/\text{x}^2.\,\text{dx} = \log_e |x^2| + c$$

$\int 1/\sin x.\, dx = \log_e |\sin x| + c$

$\int 1/e^x.\, dx = \log_e |e^x| + c$

etc.

It may be noted that the above formula works only when numerator is 1 and denominator is x or linear expression of x only as follows:

$\int 1/(x+2).\, dx = \log_e |x+2| + c$

$\int 1/(2x+3).\, dx = \frac{1}{2} \log_e |2x+3| + c$

xii. Finding ∫[f(x).g(x)] dx incorrectly

We know that,

$\int [f(x) \pm g(x)]\, dx = \int f(x)dx \pm \int g(x)dx$

But students often try to apply the same analogy to find integration of product of two functions as:

$\int [f(x).g(x)]\, dx = \int f(x)dx . \int g(x)dx$,

which is wrong.

The method for finding the integration of the product of two functions is called as Integration by parts, and it's formula is given by:

$$\int [f(x).g\,(x)]dx$$

$$= f(x)\int g(x)dx - \int \left(\frac{d}{dx}f(x)\right)\left(\int g(x)dx\right) dx$$

For example, to get ∫ x. sin x dx

Let I = ∫ x. sin x dx.Taking *x* as first function and sin *x* as the second function and integrating by parts, we obtain,

$$I = x\int \sin x.\,dx - \int\left\{\left(\tfrac{d}{dx}x\right)\int \sin x\,.\,dx\right\}.\,dx$$

$$= x\,(-\cos x) - \int 1.(-\cos x).\,dx$$

$$= -x\cos x + \sin x + C$$

xiii. Finding ∫ [f(x)/g(x)] dx incorrectly

We know that,

$$\int [f(x)\ \pm\ g\,(x)\,]\,dx = \int f(x)dx \pm \int g(x)dx$$

but students often try to apply the same analogy to find integration of quotient of two functions as :

$$\int [f(x)\,/\,g\,(x)\,]\,dx = \int f(x)dx\,/\int g(x)dx\,,$$

which is wrong.

There is no direct formula for finding the integration of quotient of two functions and the methods of finding the integration in such cases vary from case to case basis.

For evaluating

$$\int \frac{2x}{1+x^2} . dx;$$

We have to solve it by substitution.

$$Let, 1 + x^2 = t \ or$$

$$2x. \frac{dx}{dt} = 1 \ or \ 2x. dx = dt$$

The given integration reduces to:

$$\int \frac{2x}{1+x^2} . dx = \int \frac{1}{t} . dt = \log_e|t| + \mathrm{c}$$

$$= \log_e|1 + x^2| + \mathrm{c}$$

xiv. Dealing with limits within definite integration:

While attempting questions of definite

integration by way of substitution, sometimes limits are not changed to match the new variables, and that makes an error.

For example, for solving

$$\int_0^1 \frac{2x}{1+x^2}.dx$$

Let

$$1 + x^2 = t \; or \; 2x.$$

$$\frac{dx}{dt} = 1 \; or \; 2x.dx = dt$$

$$\int_0^1 \frac{2x}{1+x^2}.dx = \int_0^1 \frac{1}{t}.dt$$

$$= [\log_e|t| + \mathrm{c}\,]_0^1$$

$$= \log_e|1| - \log_e|0|$$

The solution, as stated above, is incorrect as while substituting $x^2 + 1$ as t; the new values of limits for the variable t have not been replaced.

The correct solution will be as under:

$$\int_0^1 \frac{2x}{1+x^2}.dx$$

$$Let\ 1+x^2 = t\ or\ 2x.\frac{dx}{dt} = 1\ or\ 2x.dx = dt$$

When $x = 0$, $t = 1$ and when $x = 1$, $t = 2$

$$\int_0^1 \frac{2x}{1+x^2}.dx = \int_1^2 \frac{1}{t}.dt = [\log_e|t| + \mathrm{c}\,]_1^2$$

$$= \log_e|2| - \log_e|1\,| = \log_e|2| - 0 = \log_e|2|$$

xv. Confusion over e^x and a^x:

Here *a* and *e* both are constants, and that is why the confusion exist. The constant e is a unique constant given by, $\boldsymbol{e \approx 2.71828.....}$ .The following series is used to calculate the value of e:

$$e^x = 1 + x + \frac{x^2}{2!} + \frac{x^3}{3!} + \frac{x^4}{4!} + \frac{x^5}{5!} + \cdots$$

$$\frac{d}{dx}e^x = \frac{d}{dx}\left(1 + x + \frac{x^2}{2!} + \frac{x^3}{3!} + \frac{x^4}{4!} + \frac{x^5}{5!} + \cdots\right)$$

$$= \left(0 + 1 + \frac{2x}{2!} + \frac{3x^2}{3!} + \frac{4x^3}{4!} + \frac{5x^4}{5!} + \cdots\right)$$

$$= 1 + x + \frac{x^2}{2!} + \frac{x^3}{3!} + \frac{x^4}{4!} + \frac{x^5}{5!} + \cdots = e^x$$

Here a is any constant and it may include e as well: As $\frac{d}{dx} a^x = a^x . \log_e a$,

If we apply this formula for a = e, we get,

$$\frac{d}{dx}(e^x) = e^x . \log_e e = e^x . 1 = e^x$$

Thus, the formula

$$\frac{d}{dx} e^x = e^x$$

is a special case of the formula

$$\frac{d}{dx} a^x = a^x . \log_e a$$

xvi. Double derivative of parametric functions

We know that for a parametric function, x = f (t) and y = f (t), derivative of y w.r.t. x is given by the formula,

$$\frac{dy}{dx} = \frac{dy/dt}{dx/dt}$$

However, students commit a mistake by extending the formula for getting double derivative as follows:

$$\frac{d^2y}{dx^2} = \frac{d^2\,y/dt^2}{d^2x/dt^2}$$

Example:

If $y = t^2$ and $x = t^3$. Find

$$\frac{d^2y}{dx^2}.$$

Incorrect Solution:

$dy/dt = 2t$ and

$$\frac{d^2y}{dt^2} = 2$$

$dx/dt = 3t^2$ and

$$\frac{d^2x}{dt^2} = 6t$$

Therefore,

$$\frac{d^2y}{dx^2} = \frac{d^2\,y/dt^2}{d^2x/dt^2} = \frac{2}{6t} = \frac{1}{3t}$$

Correct Solution:

dy /dt = 2t and dx / dt = 3t²

Therefore,

$$\frac{dy}{dx} = \frac{dy/dt}{dx/dt} = \frac{2t}{3t^2} = \frac{2}{3t}$$

Now,

$$\frac{d^2y}{dx^2} = \frac{d}{dx}\left(\frac{dy}{dx}\right) = \frac{d}{dx}\left(\frac{2}{3t}\right) == \frac{d}{dt}\left(\frac{2}{3t}\right).\frac{\mathrm{dt}}{\mathrm{dx}}$$

$$= \frac{\frac{d}{dt}\left(\frac{2}{3t}\right)}{\frac{\mathrm{dx}}{\mathrm{dt}}} = \frac{-\frac{2}{3t^2}}{3t^2} = \frac{-2}{9t^4}$$

xvii. Confusion Over Relation and a Function

We may define Cartesian product of two sets A and B as under:

Ax B = {(*x*, *y*):*x*∈ A and *y*∈ B}

A set R is a relation from A to B if R ⊂ A X B.

Thus, if A = {1, 2} and B = {3, 4}

A x B = {(1, 3), (1, 4), (2, 3), (2, 4)}

R = {(1, 3), (2, 3)} $\subset$ A X B and hence is a relation.

A Function is a particular Relation in which repetition of first elements of the ordered pairs never takes place.

R_1 = {(1, 3), (2, 3)} $\subset$ A X B and hence it is a relation, and its first element is unique, and therefore it is a function also. However, R_2 = {(1, 3), (1, 4)} $\subset$ A X B and accordingly is a relation but its first element one has appeared twice, and thus it is not a function.

Students should note that all functions are relations but all relations are not functions.

xviii. Wrong Meaning of Inverse of a Function

In exponents, we may write x^{-1}or reciprocal of x as 1/x, which is perfectly fine.

Students try to imitate it while writing inverse of a function as:

$f^{-1}(x) = 1/f(x)$, which is not correct.

If $f(x) = \sin x$, $f^{-1}(x) = \sin^{-1} x \neq 1/\sin x$

xix. f (x + y) = f(x) + f (y)

Students often assume that in notation of function f (x) stands for f.x and thus apply distributive law on functions incorrectly as follows:

$f(x + y) = f(x) + f(y)$, which is not true.

Thus if $f(x) = \log x$

Then, $f(x + y) = \log(x + y) \neq \log x + \log y$

Similarly, if $f(x) = x^2$

Then, $f(x + y) = (x + y)^2 \neq x^2 + y2$

xx. f (c. x) = c. f (x)

Again, $f(c.\,x) = c.\,f(x)$ is a wrong notion.

Thus if $f(x) = \log x$

Then, $f(cx) = \log(cx) \neq c.\log x$

Similarly, if $f(x) = x^2$

Then, $f(cx) = (cx)^2 \neq c.x^2$

BY THE SAME AUTHOR

You would also like to read another book **(a bestseller at www.amazon.com)** authored by me titled **"BE A HUMAN CALCULATOR."** The book is about observation based calculation tricks for improving calculation speed of the students. The book will not only help the students to do calculations at a faster speed but will also assist them in reducing their computational errors by improving their interest and creativity in mathematics.

10
Other Silly Mistakes

i. Writing 0 in Place of a Null Matrix

A Matrix in which all the elements are zero is called a Null Matrix, and its mathematical symbol is O.

However, sometimes student use 0 (zero) to denote a Matrix. Students should note that a matrix is a rectangular arrangement of numbers and hence it is not a number.

If A is any Matrix, then

A + (-A) = 0 (incorrect)

A + (-A) = O (correct)

ii. Using Division in Matrix

See the following calculation in respect of Matrices A, B

A. B = A => B = A/A = 1

The above calculation is wrong on two counts. First, the operation of division is not allowed for matrices and second, as matrices are not numbers, they cannot be canceled out as numbers.

So, the correct solution is as under:

$A.\ B = A$

$A^{-1}\ A.B = A^{-1} \cdot A$ (Multiplying both sides by A^{-1})

$I.B = I \quad (\because A^{-1}\ A = I)$

$B = I\ (\because I.B = B)$

Here, I is the Identity Matrix (a Matrix whose diagonal elements are 1 and remaining elements are zero).

iii. Matrix Multiplication is Commutative

In mathematics Matrices are denoted by capital English alphabets like A, B, C,etc., students treat them at par with Algebraic variables and assume that matrix multiplication is Commutative.

In Algebra, $x.y = y.\ x$(true)

In Arithmetic, $2.\ 3 = 3.\ 2$ (true)

However for two matrices A and B,

A. B = B. A (not necessarily true)

Example -1

Let $A = \begin{pmatrix} 2 & 3 \\ 4 & 5 \end{pmatrix}$ and $I = \begin{pmatrix} 1 & 0 \\ 0 & 1 \end{pmatrix}$

Here $A.\,I = \begin{pmatrix} 2.1 + 3.0 & 2.0 + 3.1 \\ 4.1 + 5.0 & 4.0 + 5.1 \end{pmatrix} = \begin{pmatrix} 2 & 3 \\ 4 & 5 \end{pmatrix}$

And similarly, $I.\,A = \begin{pmatrix} 2 & 3 \\ 4 & 5 \end{pmatrix}$

$\because$ A. I = I. A, A and I are commutative.

$\therefore$ The product of two matrices is always commutative when one of them is an identity matrix.

Example -2

Let $A = \begin{pmatrix} 2 & 3 \\ 4 & 5 \end{pmatrix}$ and $O = \begin{pmatrix} 0 & 0 \\ 0 & 0 \end{pmatrix}$

Here $A.\,O = \begin{pmatrix} 0 & 0 \\ 0 & 0 \end{pmatrix} = O.A$

$\because$ A. O = O. A, A, and O are commutative.

∴ The product of two matrices is always commutative when one of them is a Null matrix.

Example -3

Let $A = \begin{pmatrix} 2 & 3 \\ 4 & 5 \end{pmatrix}$ and $B = \begin{pmatrix} 4 & 1 \\ 5 & 3 \end{pmatrix}$

Here $A.B \neq B.A$

∴A and B are not commutative.

Example -4

Let $A = \begin{pmatrix} 2 & 0 \\ 0 & 5 \end{pmatrix}$ and $B = \begin{pmatrix} 4 & 0 \\ 0 & 3 \end{pmatrix}$

Here $A.B = B.A = \begin{pmatrix} 8 & 0 \\ 0 & 15 \end{pmatrix}$

∴A and B are commutative.

From the above examples, it is evident that product of two matrices may or may not be commutative.

iv. Matrix Multiplication is Associative

In Algebra, $x. (y.z) = (x.y).z$ (true)

In Arithmetic, 2. (3.5) = (2.3). 5 (true)

However for three matrices A, B, and C;

A .(B.C) = (A. B). C (not necessarily true)

The Associative Law holds good in matrix multiplication only when B. C and A. B exist. That means, the number of columns in matrix B should be equal to the number of rows in matrix C and number of Column in the matrix A should be equal to the number of rows in matrix B.

v. Product of two matrices is O only when at least one of the matrix is O

A. B = O is true when A = O

A. B = O is true when B = O

A. B = O is true when A = O and B = O

However, A. B = O even if none of A and B is a null matrix. See the example given on next page:

Let A $=\begin{pmatrix} 2 & 1 \\ 4 & 2 \end{pmatrix}$ and B $=\begin{pmatrix} 1 & 0 \\ -2 & 0 \end{pmatrix}$

$$A.\,B = \begin{pmatrix} 2.1 + 1.(-2) & 2.0 + 1.0 \\ 4.1 + 2.(-2) & 4.0 + 1.0 \end{pmatrix} = \begin{pmatrix} 0 & 0 \\ 0 & 0 \end{pmatrix}$$

Hence, the product of two matrices can be a null matrix even if none of the matrices is a null matrix.

vi. Using Algebraic Identities on Matrices

As explained on previous pages, students try to apply laws of algebra on matrices. However, applying identities of algebra on matrices is incorrect.

We know that,

$(a + b)^2 = a^2 + b^2 + 2a.b$, holds good for Algebra.

Now let us explore for two matrices A and B,

$(A + B)^2 = (A + B).(A + B)$

$= A.A + A.B + B.A + B.B$

$= A^2 + A.B + B.A + B^2$

And we had seen that in matrix, A. B may or may not be equal to B. A and thus for matrices,

$(A + B)^2 = A^2 + B^2 + 2A.B$ will hold good only when A. B = B. A i.e. the matrices are commutative.

Given the above discussion, students are advised not to use algebraic identities on matrices without knowing the conditions involved.

vii. Writing 0 in place of a null vector

Sometimes, 0 is used to represent null vector whose symbol is $\vec{0}$.

Thus,

$\vec{a}.\vec{b} = 0$ is correct as it is a scalar product of two vectors in which answer is a scalar.

However,

$\vec{a} \times \vec{b} = 0$ is incorrect as it is a vector product of two vectors in which answer is a vector. We should write it as $\vec{a} \times \vec{b} = \vec{0}$.

viii. Using algebraic identities on vectors

We had seen in the previous topics that we can not apply algebraic identities to the matrices.

However, as the scalar product of two vectors is commutative and is a scalar, the algebraic identities hold good for vector scalar product. Thus, it is not wrong to say that:

$(\overrightarrow{a} \pm \overrightarrow{b})^2 = \vec{a}^2 + \vec{b}^2 \pm 2\vec{a}.\vec{b}$

However, students should note that,

$(\overrightarrow{a} \pm \overrightarrow{b}) \text{ x } (\overrightarrow{a} \pm \overrightarrow{b}) \neq (\vec{a} \pm \overrightarrow{b})^2$

As such, algebraic identities may not work for cross-product of vectors.

ix. Confusion over "Or" and "Nor."

In Set Theory and Probability, the symbols ∪(Union of two sets) and ∩ (Intersection of two sets) are used to denote the meaning of "Or" and "And" respectively.

Thus, in Probability P $(A \cup B)$ and P $(A \cap B)$ denotes Probability (Event A or Event B) and Probability (Event A and Event B) respectively.

However, students use the word "Nor" in the sense of "Or" whereas the meaning of "Nor" is "End."

Thus, Probability (neither A nor B)should be represented by the symbol P $(A^c \cap B^c)$ and not P $(A^c \cup B^{c)}$as usually understood by the students.

x. Confusion over "Experiment" and "Event."

Students are often confused in two terms "Experiment" and "Event" in Probability.

By Experiment means some activity which is done in anticipation of a result, whereas Events are outcomes or results or observations of the event.

For Example:

1. Tossing a coin is an Experiment. Getting head or tail, while tossing a coin are examples of events.
2. Throwing a dice is an experiment. However, "getting an even number," "getting an odd number," "getting a prime number" are the different events related to this particular experiment.

xi. Wrong use of Section Formula

Section Formula: For two endpoints of line segment $A(x_1, y_1)$ and $B(x_2, y_2)$, the coordinates of the point P(x, y) that divides the given line segment in the ratio m: n internally is provided by:

$$\left(\frac{m\,x_2 + n\,x_1}{m+n}, \frac{m\,y_2 + n\,y_1}{m+n}\right)$$

If point P(x, y) divides the given line segment in the ratio m: n externally, then its coordinates is given by:

$$\left(\frac{m\,x_2 - n\,x_1}{m-n}, \frac{m\,y_2 - n\,y_1}{m-n}\right)$$

The diagram in the case of internal division is as under:

A----------------P----------------B

Where AP: PB = m: n

However, students are of the view that the diagram of external division works as follows:

A----------------B----------------P

Where AB: PB = m: n (wrong assumption)

For external division also,

Where AP: PB = m: n (correct ratio to apply in the formula).

xii. Confusion over Permutation or Combination

There is another widespread confusion among students, and that is the difference between Permutation and Combination.

Permutation and Combination are two closely related concepts. The underlying meaning of Permutation is "Arrangement," and that of the combination is "Selection."

From the word 'Combination,' we get an idea of 'Selecting several objects out of a large group.'

Example: If there are three persons A, B, and C, in how many ways we can choose two individuals out of them?

The above example is a case of "Combination," and the possible number of ways are three (AB, BC, CA). In Combination, the order of objects is of no importance. Thus AB and BA convey the same meaning.

On the other hand 'Permutation' is all about "Arranging several objects in different orders out of a large group."

Example: A group of 3 students A, B, and C is getting ready to take a photo for their annual gathering. In how many ways they can sit for the photograph?

The above example is a case of "Permutations," and possible ways are ABC, ACB, BAC, BCA, CAB, and CBA. In Permutations, the order of objects has its significance, and hence AB and BA are considered as different Permutations.

xiii. 0! = 0

Explanation-1

n! is defined as the product of all positive integers from 1 to n.

Therefore, n! = 1.2.3.4……… (n-1).n

Then:

1! = 1

2! = 1.2 = 2

3! = 1.2.3 = 6

4! = 1.2.3.4 = 24

and so on.

n! can also be expressed n.(n-1)! .

For n=1, using n! = n. (n-1)!

We get 1! = 1.0!

This yields, $0! = 1$

Explanation-2

The idea of the factorial is used to compute the number of permutations of arranging a set of n objects. The number of permutations (number of ways of arranging objects) of n objects taken all at a time = n! Thus, observe the table on the next page:

No and names of the objects (n)	Number of Permutations (n!)	List of Permutations
1 (A)	1	{A}
2 (A,B)	2	{(A,B), (B,A)}
3 (A,B,C)	6	{(A,B,C), (A,C,B), (B,A,C), (B,C,A), (C,A,B), (C,B,A)}
0	1	{ }

Therefore,

We may form an empty set only in one way, so 0! = 1.

Explanation-3

We know that,

^{n}Cr describes the number of ways of selecting r persons out of n persons.

We also know that,

$^{n}C_{r} = n! / r! (n - r)!$ ------ (1)

From (1), we have,

$^{n}C_{n} = n! / n! (n - n)! = 1/ 0!$ ----- (2)

We also have, $^{n}C_{n} = 1$, as number of ways of selecting n persons out of n persons is 1. ---- (3)

From (2) and (3), we have

$1/ 0! = 1$

This yields, 0! = 1

xiv. Surface of a 3-D Object

Students are always confused about the term Surface of a 3-D object.

The surface of a 3-D object means, outer or inner part of an object, we can touch or see. In other words, any part of the object, which is in contact with air, may be termed as Surface of that object.

xv. Surface Area vs. Total Surface Area

Similarly, I have always found students confused about the terms Surface Area and Total Surface Area.

In fact, there is no difference between these two terms. The term surface area means the total surface area of the object.

xvi. Circumference = Perimeter

The perimeter of a 2-D figure is the length of the outer boundary of a 2-D figure.

Circumference, on the other hand, may be defined as the length of the outer curved boundary of a 2-D figure.

In a circle, the entire outer boundary is curved, and hence we may say that its perimeter is equal to its circumference.

Not knowing the correct definitions, may invite trouble for students in some cases as explained in the following example:

What is the perimeter and circumference of a semi-circle?

A student, who doesn't know these definitions, may end up getting the answer of both as $\pi.r$ or $\pi.r + 2r$.

However, the student who knows the correct definition will get the answers as follows:

Perimeter = $\pi.r + 2r$ (the entire outer boundary)

Circumference = $\pi.r$ (the outer curved boundary)

xvii. Volume ≠ Capacity

Students are always confused about these two terms and use them interchangeably for each other, which is not correct. Both these terms have a specific meaning and students should use these terms accordingly.

Go through the following example:

Find the capacity of a cylindrical glass whose radius is 7 cm and whose height is 20 cm.

Capacity of Glass = Volume of Glass

$$= \pi . r^2 . h$$

$$= \frac{22}{7} \times 7 \times 7 \times 20$$

$$= 3080 \text{ cubic cm}$$

Students should under that Volume of a 3-D object is the amount of 3-D space occupied by that object whereas Capacity is the maximum quantum of material (solid, liquid or gas) which can be held by that object.

Students will understand the difference through the example of a hollow cylindrical glass as shown in the diagram shown on next page. The Outer radius of the Glass is R, and inner radius is r. The height of the glass is h.

Now the Volume of the Glass = Volume of material required to make it = Volume of outer Cylinder – Volume of inner Cylinder = $\pi.R^2.h$ $\pi.r^2.h = \pi.(R^2 - r^2).h$

Whereas the capacity of glass = Maximum Quantity of water held by the glass = Volume of inner cylinder = $\pi.r^2.h$

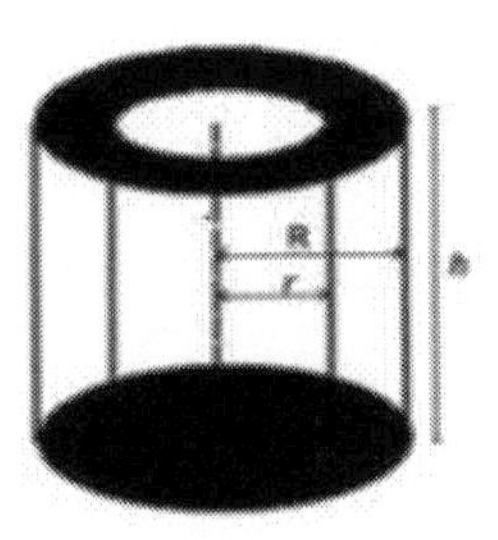

Students may see that both the terms are entirely different.

Then, what is wrong in the example given on the per-page.

The correct solution is as under:

Capacity of Glass = Volume of water in the glass

$$= \pi.r^2.h = \frac{22}{7} \times 7 \times 7 \times 20$$

$$= 3080 \text{ cubic cm}$$

Keep some water in a cylindrical glass at a freezing temperature. The ice in the glass is taken out.The ice will now look like a solid cylinder, and hence the volume of the water (which is now ice) will be equal to the volume of a solid cylinder, i.e., $\pi.r^2.h$.

xviii. Confusion over Modulus of a number

We know that $|x|$ represents the absolute value of a number. For example:

$|7| = 7; |0| = 0; |-5| = 5$

The definition of $|x|$ is:

$|x| = x$, if $x \geq 0$ and $|x| = -x$, if $x < 0$

Students are often confused with the definition of $|x|$. They are of the view that if $|x|$ always returns the positive value as the answer, then how

$|x| = -x$.

As per definition of $|x|$, the function works as under:

$|7| = 7$

$|-5| = -(-5) = 5$

That means for $x \geq 0$, $|x|$ returns the same number as the answer and hence its definition is

$|x| = x$, if $x \geq 0$

And for a negative number, $|x|$ returns the negation of that negative number i.e. again a positive number and hence its definition is:

$|x| = -x$, if $x < 0$

A WORD OF THANKS

Dear Readers,

Thank you for reading this book!

If you enjoyed it or found it useful, kindly post a short book review on Amazon via https://www.amazon.com/Avoid-Mistakes-Mathematics-Rajesh-Sarswat/dp/1520443978/

Your support will make a difference to me by making this book even better.

Thanks again for your support.

RAJESH SARSWAT

Made in the USA
Middletown, DE
08 August 2018